SEA URCHINS

ANIMALS WITHOUT BONES

Jason Cooper

Rourke Publications, Inc.
Vero Beach, Florida 32964

PHOTO CREDITS
© Lynn M. Stone: cover, title page, pages 4, 7, 10, 12, 15, 17, 21;
© James P. Rowan: page 8; © Herb Segars: page 13; © Frank
Balthis: page 18

Library of Congress Cataloging-in-Publication Data
Cooper, Jason, 1942-
    Sea urchins / by Jason Cooper.
        p. cm. — (Animals without bones)
    Includes index.
    Summary: A simple introduction to the physical characteristics,
life cycle, and habitat of sea urchins and related species.
    ISBN 0-86625-570-2
    1. Sea urchins—Juvenile literature. [1. Sea urchins.] I. Title.
II. Series: Cooper, Jason, 1942- Animals without bones.
QL384.E2C66 1996
593.9'5—dc20   95-26009
                                   CIP
                                   AC

Printed in the USA

TABLE OF CONTENTS

SEA URCHINS

Sea urchins are called the pincushions of the sea — and for a good reason. Many of the ball-shaped sea urchins are full of sharp, needlelike spines.

Each spine stands up straight, but it can change direction, like your fingers.

The soft parts of a sea urchin are protected by hard, thin plates. Together, these plates are the urchin's outer skeleton, or **test** (TEST).

The hard, lightweight plates that make up an urchin's test show up well after the animal is dead and dry

WHAT SEA URCHINS LOOK LIKE

You might never guess that sea urchins and sand dollars are very close cousins. Sand dollars — some are called keyhole urchins — are flat as cookies.

Sand dollars have spines, but they're very small. Sand dollars are covered with a carpet of tiny, hairlike bristles.

Urchins have dozens of tiny, tubed feet along with their spines. The feet help them stick to things under water.

A sea urchin's mouth is hidden on the underside of its body.

Cookie-shaped sand dollars are sea urchins with short, hairlike spines

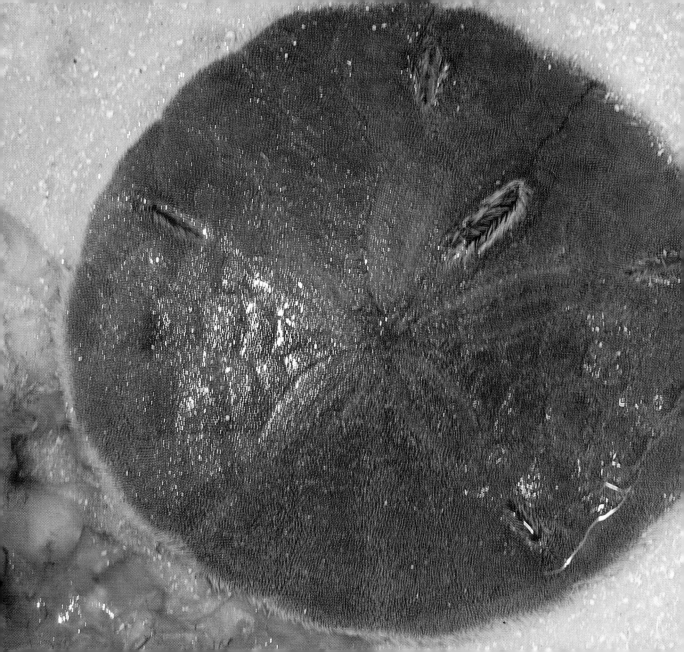

KINDS OF SEA URCHINS

Hundreds of **species** (SPEE sheez), or kinds, of sea urchins live in the world's oceans. Sand dollars and plump sea biscuits are sea urchins with different shapes.

Some kinds of urchins have much longer and sharper spines than others. The spines of long-spined black urchins are sharp as needles. Pencil urchins have shorter, duller spines.

The shingle urchin has flat spines that look like canoe paddles.

Shingle urchins have flat spines shaped like canoe paddles

THE SEA URCHIN FAMILY

Sea urchins and their cousins belong to a group of small, boneless animals called **echinoderms** (ee KIY no dermz). All echinoderms are **marine** (muh REEN), or sea, creatures.

Sea stars, sometimes called starfish, are the best known of the echinoderms.

A sea star isn't really much different than a sea urchin. If a sea star's arms were folded up to meet at their tips, it would look much like a spineless urchin.

Green urchins and a sea star share a tide pool in Maine

The long, needlelike spines on a black urchin keep most predators away

A young diver uses gloves to handle this long-spined urchin

WHERE SEA URCHINS LIVE

Urchins live on different kinds of ocean bottoms all over the world, often close to shore. Urchins live in the warm, tropical waters of the Caribbean Sea, for example, as well as the icy water of the Antarctic Ocean.

Spiny urchins often live on rocks. Many can bore, or drill, holes into the rock. The holes become urchin homes.

Other species live in sea grass. These urchins **camouflage** (KAM uh flahj), or hide, themselves. They are well hidden in the grass jungle because they attach bits of seaweed and shell to their tests.

Sand dollars have sharp edges so they can burrow, or dig, into sand.

A sea urchin in Florida hides under a camouflage coat of shells

BABY URCHINS

Sea urchins carry hundreds of thousands of eggs in their tests. An urchin releases eggs into the sea. A tiny **larva** (LAR vuh) hatches from an egg.

The larva is a baby urchin, but it doesn't look like its parents.

The urchin larvas go through many changes as they grow. Only a few grow up to be adult urchins. Most larvas will end up as food for a fish or some other creature.

Urchins that survive to become adults live in groups called colonies

HOW URCHINS LIVE

Sea urchins are not very active. When they do move, they use their spines and tubed feet to go slowly from place to place. Thousands of urchins or sand dollars often share the same neighborhood. These huge groups are called colonies.

A colony of hungry urchins can gobble up nearly all the marine plants in a small area.

Tiny fish often find safety by living among the urchins' spines.

Urchins can walk on their spines, like skinny stilts

PREDATOR AND PREY

Sea urchins live mostly on a diet of sea plants. An urchin chews up food — and sometimes rock — with five "teeth" in its mouth.

Some urchins eat food that drifts into their spines and tubed feet. The spines help trap the food.

In turn, sea urchins are **prey** (PRAY), or food, for such **predators** (PRED uh torz) as lobsters, fish, and sea otters. Without sea otters to eat them, spiny sea urchins would destroy too many sea plants in the North Pacific Ocean.

Urchins have "teeth" around their mouths (center of photo) to grind food

PEOPLE AND SEA URCHINS

Sea urchins can be tasty, and they can be dangerous.

In several parts of the world, including Japan, people eat urchins. Urchin eggs are sold on the streets of Barbados.

Divers and swimmers are careful with spiny urchins. Urchins can't throw their spines. A person has to touch a spine to be hurt by it. The spines of some species are poisonous. Even nonpoisonous spines can cause bloody wounds.

Most urchin spines usually hurt about as much as a bee sting.

Glossary

camouflage (KAM uh flahj) — coloring that allows an animal to blend into its surroundings

echinoderm (ee KIY no derm) — a group of small, boneless sea animals, including urchins, sea stars, sea cucumbers, and others

larva (LAR vuh) — an early stage of life in some animals; the young animal does not look like the adult it will become

marine (muh REEN) — of or relating to the sea and salt water

predator (PRED uh tor) — an animal that kills other animals for food

prey (PRAY) — an animal that is killed by another animal for food

species (SPEE sheez) — within a group of closely-related animals, one certain kind, such as a *long-spined black* urchin

test (TEST) — the hard, platelike sections that cover the soft, inner parts of urchins and their cousins

INDEX